THE NIGHT FIGHTER WORLD

THE NIGHT FIGHTER WORLD

Remembering My Time in the Royal Air Force (1940–1946)

Peter L. Croft

Book Guild Publishing
Sussex, England

First published in Great Britain in 2011 by
The Book Guild Ltd
Pavilion View
19 New Road
Brighton, BN1 1UF

Typesetting in Garamond by
Keyboard Services, Luton, Bedfordshire

Printed in Great Britain by
CPI Antony Rowe

A catalogue record for this book is available from
The British Library

ISBN 978 1 84624 580 0

'It was a fine life as only those who have served on an RAF squadron would know.'

Acknowledgements

Without the contributions of the following, these memoirs could not have been produced:

The late Cameron Cox (for some details in his memoirs).

Roger Kirkman (who gave advice on radar and other matters).

Michael J.F. Bowyer (quotes from his book *Air Raid*).

Andrew Thomas (quotes from his book *Squadrons of the RAF*).

The late C.F. Rawnsley and Robert Wright (quotes from their book *Night Fighter*).

J.D.R. Rawlings (quotes from his book *Fighter Squadrons of the RAF and their Aircraft*).

The Public Record Office, Kew.

Chapter I

'Beware of memory, even one's own.'
(Sir Martin Gilbert, historian)

About the middle of 1940 I received notice of call-up for the services; my previous unsuccessful attempts in 1938 to join the Royal Air Force as a pilot were completely discounted. I reported to Chatham Drill Hall not knowing which service I was to join; there were several hundred of us. You must believe me when I explain that it was sheer luck that I got into the RAF at all; well, here is my version of what happened at that most important first call-up.

We were positioned in line two or three deep, and then an army officer appeared accompanied by two sergeants. He commenced at one end of the line, and after counting about fifty men, placed a sergeant in charge to usher them into another room. Afterwards we heard they were destined for the navy. He came along the line again and counted off the next lot, who were told to remain where they were for the army. Yes, you've guessed it, I happened to be standing with the remaining group of about fifty or so, and we were taken into another side room to be registered for the RAF. Oh boy, was I relieved at that bit of luck.

Two months after my experience at Chatham I received papers, and a travel warrant, with instructions to report to RAF Cardington together with small kit (we were to be there for two days). First there was a medical, and I presumed I passed as, generally, you are only told if you do not. I pleaded with the doctor about my eyes, and he was sympathetic, but adamant that I would never pass for aircrew. Then came an interview, before two civilians, and the questions were about my life, education and job. The next day's experience was more formal when I had to complete a written examination, and then had to appear before another selection board who were in uniform; needless to say, there was no indication of what was in store for me.

I had commenced working for SAWFA, the South African Wine Farmers Association (London) Ltd in 1938, and after eighteen months in the sample room had since been clerking in the bonded warehouse, and gaining much experience in the handling of the importation of casks of wine. It was late November before I received more news about call up: this was an instruction to report to RAF Blackpool, and on 20th December, 1940, I left for the war.

We were living in Sevenoaks at the time, and I well remember that we had experienced several nights of enemy air activity, including one raid when a shower of small incendiary bombs landed on our front doorstep and garden, and we kicked them out into the road. I expect I was jolly glad to leave that area, and report to peaceful Blackpool.

The nicest decision ever taken by SAWFA was to pay an allowance to its employees during the time they were away in the forces. We were paid half our salaries at time of call-up with no deductions in lieu of service pay. In my case it was £2 per week: £1 paid into my bank account every week, and £1 kept by SAWFA, as an incentive to return to work for them when demobbed. As an aircraftsman

2nd class on pay of fourteen shillings per week, I found the extra £1 most useful.

Anyway, there I was in Blackpool, Aircraftsman 2nd Class, Croft, P.L. number 1190532, doing my three weeks of square bashing, and I tell you getting used to using boots for the first time was not easy. We were billeted in what had been a boarding house in peace time, quite comfortable but with an inadequate amount of food. We had enormous appetites, being out most of the day in the breezy Blackpool air, drilling, marching or doing physical exercises, and we used any spare cash to buy extra food.

I was kindly invited to spend Christmas Day with Jim Fox and family. Jim was a cousin many times removed, and very hospitable. It was unfortunate that the army chaps billeted with him had experienced terrible times during the evacuation from Dunkirk. The fact that RAF fighters had apparently rarely been seen over the area to defend them was very much in their minds, and they made sure I was informed of this. Having been in the service for only five days, I had not been made aware of the circumstances there, and had no defence.

Who should I meet in Jim's house but first cousin Cecil Fox. Together we had attempted to join the RAF in 1938, and he had been successful, but not as a day fighter pilot as we had intended; because he was over twenty-four years old he had been passed on to coastal Command. He was already in possession of his 'wings' and sergeant's stripes, and was on a special wireless course in Blackpool. We mere aircraftsmen were confined to billets after 10p.m., and with service police there in plenty, one did not dare to disobey the normal rules. However, I did go out on the town a couple of times with Cecil wearing his spare sergeant's tunic, and thereby avoiding arrest. My landlady was very nice and did not report me, as she should have done, for coming in late.

My posting from Blackpool was to No. 3 Maintenance Unit at RAF Didcot, near Oxford. There we dug the ground, heaved things about, and generally acted as labourers, apparently awaiting a further posting. I do remember that the chaps there with me were similar to me in class, no toffs, and no working class, if you know what I mean, just middle class. I suppose the interviews we had attended at Cardington had sorted us out. I flatter myself by suggesting we were slightly more intelligent than average.

After about two months at Didcot I received my proper posting; it was to RAF Worth Matravers, near to Swanage, on the Dorset coast. It was a lovely place to be sent to, and I enjoyed my short time there. Worth was a radar station, in the front line to pick up any enemy air activity approaching the UK. We worked in shifts for twenty-four hour coverage, but for the time I was there I was under training, and was supervised when it was my turn to watch the radar screen. It was my first meeting with the Women's Auxiliary Air Force: two thirds of the crews, and most of the NCOs were WAAF, and I was most impressed by their professional approach to the work. In the spring of 1941 there was very little enemy activity in our sector, and we spent our time tracking friendly aircraft. Most of the action was further east, where London was again getting a hammering.

My billet in Swanage was above a wine shop, privately owned. The owner, with little to sell because of the restriction on imports, was having a difficult time financially, and that is why he took in lodgers. Several of us were comfortably billeted there. This included Raymond

Jacquerello, a former actor in minor parts, and with a knowledge of life that I had not. We had very little income to finance our social activities in Swanage, and over the ferry into Bournemouth, and normally spent our pay, and my extra £1 per week, long before the next pay day.

I had only been at Worth for a couple of months when an RAF order appeared on the notice board stating that any ground radar operator could volunteer for a new aircrew trade as an air radar operator. Most importantly it went on to say that although in all other respects the applicants had to reach the full standard of fitness for flying, if they had only a slight eyesight problem, and wore the necessary correcting lenses, they could apply for selection. I applied at once, and when I tell you that within a week or so I was called forward for a selection board and medical, believe me it must have meant the RAF had need of something very urgent; normally remustering in the services is painfully slow.

With hindsight I can tell you that the strategic plans of the Royal Air Force had been completely disrupted by the German Air Force change of tactics in the autumn of 1940, after the end of the Battle of Britain in which they had been defeated by our day fighters. When the Luftwaffe took to night bombing, the RAF was unprepared for large scale defence at night. Something had to be done quickly to protect the cities, the ports, the factories and the people. The devastating air raid on Coventry on the night of 19th November, 1940 may well have had as much effect politically as the London blitz. In the winter of 1940–1941 there is no doubt that a state of panic existed in the higher command of the RAF at the situation which confronted them. Although there was an urgent need for an increase in all aspects of night defence, for some time it had been thought that the night fighter aeroplane would be the most effective. Of course until then priority had been given to the production of day fighters and the training of their pilots.

This is not to say that defence by night fighters did not exist, but the most available aeroplanes were the Blenheim and the Defiant, both already failed day fighters. Because they were too slow to catch up with the enemy Junkers and Heinkels, and had little forward facing fire power, many opportunities were lost, and in any event there were few enough of them. It is recorded that in December 1940 even old Hampden bombers from Nos. 83 and 144 Squadrons, with their turrets manned, and extra beam guns fitted, were airborne as night fighters, which is an indication of the desperate measures attempted because of the nightly raids. They were, of course, no faster than the enemy bombers they sought to pursue.

Day fighter Hurricanes were also co-opted to join in night defence to get over the low speed problem, but they had to rely on searchlight cooperation, or just patrolling over a suspected target in the hope of finding a raider. This did work out sometimes because during one of the raids on Hull, No. 255 Squadron, flying Hurricanes and Defiants, shot down seven enemy bombers in one night. Another example was Flight Lieutenant Richard Stevens of No. 151, at that time ostensibly a day fighter squadron, who destroyed fourteen enemy aircraft at night over a long period in his Hurricane.

These brave efforts, however, were only a small contribution to attempt to solve the problem of the nightly raids. The frustration of the people involved in night defence at that time must have been absolute when at the height of the London blitz, from 18th to 22nd December,

1940, over 160 fighter sorties were flown at night with only one interception and no kills. On 29th December, 1940, the night of the great fire raid on London, Fighter Command instructed the fighters to fly 'layered' patrols over the city, and, to allow them freedom, the anti-aircraft guns were ordered to cease firing, but there were no interceptions.

Controlling fighters from the ground at night had not been a success so far because, although the original Chain Home early warning ground radar, protecting the east and the south of England, operating on long wavelengths with fixed aerials, was adequate for interceptions in daylight, it was not accurate enough for night operations when the pilot could only see his target at a much closer range.

Gradually this was being supplemented by GCI (Ground Control Interception) stations with rotating aerials on shorter wavelengths, which also gave more accurate height information. However, the precision in synchronising individual height, course and speed, essential to night interception, demanded specialist techniques from the ground controllers, which took time to develop. It is said that the first really successful night interception from the ground did not take place until 20th February, 1941. This was by the famous Squadron Leader Brown, known as the 'Sultan of Sopley', who arguably became the most successful ground controller of the war.

Air Interception radar, known as AI, had been in the experimental stage for some time by the autumn of 1940. The RAF Interception Unit had attached Blenheim aeroplanes with radar on board to Nos. 25 and 29 Squadrons, but in these aeroplanes when they attempted an interception they had very little chance to catch up with the target. What was needed was a fast twin-engine aeroplane with forward-firing guns, and room for another crew member to work the radar set. For whatever reason the Bristol Beaufighter had been originally designed, it was perfect for this purpose.

No. 604 Squadron commenced to convert to Beaufighters with Mark IV radar as early as July 1940, and No. 219 also flew Beaufighters with radar from October 1940. The first radar sets had their share of teething troubles, sometimes working beautifully on the ground, but registering nothing when switched on in the air. Often the vibrations of aircraft engines would be enough to knock out these sensitive sets. It was not until 20th November, 1940 that an AI-equipped aeroplane (a Beaufighter of No. 604 Squadron) first shot down an enemy bomber at night (actually a Junkers 88), after a radar interception. This battle is described in that excellent book *Night Fighter* by C.F. Rawnsley and Robert Wright.

In the winter of 1940–1941 it was essential for the RAF to put matters right quickly, and oppose the Luftwaffe successfully in the night skies over Britain before there were more casualties and more damage was done, and before more questions were asked about its defence plans. The important factors were to have better ground control in the early stages of an interception at night, improve the performance of AI, provide greater numbers of the speedy Beaufighter aeroplane, and have more trained crews.

In early 1941, with the Bristol Beaufighters coming off the production line in increasing numbers (there was the Mark II with the Rolls-Royce in-line Merlin engines, and the Mark III with Bristol Hercules radial engines), and with the airborne radar Mark IV at last improving in performance in the air, things were looking up. One must say that the forming of new operational squadrons of night fighters was then conducted at an incredibly fast pace. I guess there had been nothing like it since the day fighter explosion in the early days of the Battle of Britain. If it was not completely new units such as No. 68 Squadron, others such as No. 255 Squadron, flying Defiants, were hastily converted to Beaufighters.

So what about the crews? Take the example of my future pilot Cameron Cox. At the time he was in the final stages of training as a day fighter pilot, in fact doing the gunnery course at Sutton Bridge, when the course members were informed that they were not being posted to day fighter squadrons after all. Instead, and without any further training for night flying, they were to be sent to night fighter squadrons. Cameron found himself flying Defiant aeroplanes at night, and was not at all pleased as he had expected to be flying Spitfires or Hurricanes on a day fighter squadron. Choice was out of the question; it was an emergency, and therefore bad luck to be just ready for squadron duties at that time.

There was also the need for aircrew to man the airborne radar sets, a completely new RAF trade. It was the obvious thing to do to ask for volunteers amongst the ground radar operators because at least they knew something about radar, which would hasten training. Also the RAF must have known, from trawling through its records, that some of these ground operators had previously volunteered for aircrew, and were of the right selection, but had failed the medical only because of short sight. They were mainly looking for men to man a radar set in an aeroplane, so what was the problem if they were fully fit otherwise? Later on, when this avenue of recruitment was exhausted, and even more radar operators were still needed, the RAF extended the call to other ground crew trades.

9

You might well ask why the RAF did not take navigators, airgunners and other already trained aircrew to cover for this emergency. The answer is that Bomber Command was suffering from a severe shortage of crews due to the losses experienced in the first year of the war, and could not be expected to release any trained crews at all. Coastal Command also had none to spare. When the Defiant squadrons were converting to using the Beaufighter aeroplane, it was attempted to teach the mysteries of air radar interception to the airgunners already on the squadrons, but it was with mixed results, and many airgunners preferred to remain in the trade they were trained for, and were posted to other commands.

Anyway, the unexpected change in tactics to night bombing by the Luftwaffe might have been most serious for the country, and for the Royal Air Force, but it allowed me entry into aircrew and flying, which I had wanted to be part of for so long, and for this I am most grateful.

I had hardly returned to Swanage after my selection board interview and medical before I was posted to the newly opened No. 3 Radio School at Prestwick to take the airborne radar course (there had been no facilities for training air radar operators before this). It was early July 1941, and this may have been the first course. Anyway talk about speed – it was only a month or so previously that I had seemingly been stuck in a ground job for the rest of the war. The change of my trade from ground to air had been breathtakingly quick, and so unusual in the services.

Not only this but the course was a mere three weeks long. It consisted of lectures, demonstrations on a ground radar set, and, in my case, one flight only in an old Blenheim I aeroplane fitted with a Mark III radar set that did not work! Then we were posted direct to squadrons – no operational training units for us. Corners had been cut with no thought for anything else but to get together as many night fighter squadrons as possible for the air defence of Great Britain at night.

Chapter II

'Far too many writers rely on the classic formula of a beginning, a muddle, and an end.'

(Philip Larkin)

Now Leading Aircraftsman Croft, I was posted direct to No. 255 Squadron, stationed at Hibaldstow in Lincolnshire, together with five others from the course. Of these, Buddy Biggs, Denis Pollard, Jack Talbot and I remained friends and went about together; Denis and I continued our friendship for life. When we arrived at the squadron our group did not have the faintest idea what to do in respect of practical airmanship, and not much more about airborne radar!

Luckily, the pilots of No. 255 Squadron were converting from the single engine Defiant to the twin engine Beaufighter aeroplane, and because they were very busy, it gave us time to find out what to do before being called on to fly, and to operate an airborne radar set. It was ironic that the pilots, thinking we six were the experts having been on a course, all wanted us to crew up with them, and would have abandoned their old airgunners who were attempting to learn about radar at the squadron, helped by the ground radar officer. Anyway, in spite of the difficulties of arriving at the squadron inadequately trained, it was with great pride and joy that I had my half-wing flying brevet sewn on to my uniform tunic and

13

battledress blouse; I was aircrew at last. The lettering on the brevet was RO for radio operator (the word 'radar' was not used as it was top secret) soon to be changed to N for navigator.

During the first couple of months on 255 Squadron I flew with several pilots, but particularly with Phil Kendall and Johnny Wright. I suppose, unknowingly, we crew were quite brave because the Beaufighter twin-engine aeroplane was new to these pilots, and the Beaufighter II with Merlin in-line engines was very difficult to handle. Because of this, the squadron was non-operational for some time. Now Sergeant Croft, it was not until 18th September before I undertook my first night flight (I was thrilled to bits by this), and then the squadron was pronounced operational, and I completed my first operational patrol on 3rd October, then flying with Johnny Wright.

It is time to tell you what we air radar operators actually did. The job, in the simplest of terms, was to follow another aircraft at night by instructing the pilot accordingly. The normal procedure, whether already on patrol or scrambled from the ground, was to receive initial instruction from a controller at a GCI station. If the controller required us to investigate a possible enemy aircraft he would instruct us ground to-air by giving a series of courses to fly, which should take us as near as possible behind the target. The better the controller, the nearer and the better position in respect of the target we ended up with. However, this was not usually near enough for the pilot to see the target and therefore the radar operator took over.

The Mark IV AI radar set consisted of two small screens of cathode ray tubes set side by side. One showed the horizontal and the other the elevation position of the target. The target would show up as a 'blip' on the screens, and, ideally by giving instructions to the pilot one kept the blips steady and equal distance each side of the centre trace, which was, in effect, the nose of our aeroplane. By the blip's distance from one end of the trace one could also give the range of the target ahead, and could instruct the pilot to increase or decrease speed as required. The extent of the maximum range depended on the height of our aeroplane from the ground, because as well as picking up the signal from the target, the set would also show other detail such as the ground itself. Therefore, the lower the height of the aeroplane from the ground, the less range was available because the blip could not be seen through ground returns.

It all sounds so simple and easy to operate as I describe it above, but I can assure you it was not so. The picture on both screens was not clear cut as you might expect like a television set, but instead was a shivering confusion of wavy lines. An object like a Christmas tree blotted out part of the tubes. This was the ground returns, and on both sides of the wobbly trace there was a sort of fringe. From this very indistinct picture and background of what looked like cut grass, one had to pick out the target blip. I tell you it was not easy.

The other main issue was that enemy bomber pilots did not fly straight and level. If they had any experience at all they would weave and turn and alter height constantly in order to try to avoid being followed, and if they suspected a fighter was on their tail they would violently alter course and height all over the sky. All this had to be followed by the radar operator with his target blips hardly discernible from the background of other stuff on the screens. One of the worst errors the operator had to control was over-correcting, which meant that in a turn before the blip had centred on the trace, an instruction had to be given at the correct time to stop the turn, otherwise one's pilot would be constantly altering course, and consequently getting more irritable. Anyway, most of these difficulties were overcome by experience, which we gained by practice interceptions at every opportunity, day and night.

We radar operators who became operational in the early days of the panic caused by the Luftwaffe change of plans (as previously described) took to the air with very little, if any, experience of navigation. This knowledge we picked up on the squadron, and in my case particularly in the training period prior to going overseas. We were helped by signal beacons on some RAF bases, so that we could use our screens to guide our pilots back home. Also the ground controllers usually still had control of us if by chance the radar became unserviceable.

The Beaufighter was a most comfortable aeroplane for the crew. Unusually there was a form of central heating for the navigator seated in front of his radar set in the centre of the aeroplane. A pipe gave out hot air, obviously from engine warmth, which was most acceptable on a winter's night at 10,000 feet up in the sky. The bad news was that in the Beaufighter II aeroplane the navigator, if necessary, had to reload the guns. This might sound simple enough, but in the confined space of a small aeroplane's hull, hampered by oxygen and communication wires attached to one's helmet, lifting sixty pounds of ammunition drums and placing them correctly, after removing the empties, was exhausting. Bad enough if your pilot was flying straight and level, but much worse of course if he was following a weaving target. How many times did one hit one's head on the signal pistol protruding from the roof?

Chapter III

'Historians relate not so much as what was done as what they would have believed.'

(Benjamin Franklin)

In spite of what I said earlier both Phil Kendall and Johnny Wright took back their former airgunners to fly with them as radar operators. They had by now been trained to work a set, and this left me pilotless, and as a 'spare' I had to fly with all sorts. Then I was approached by Sergeant Cameron Cox, and after I had been tested by him in both day and night interceptions, he asked me if I would fly with him as his permanent radar operator. I cannot express enough admiration for, and how lucky I was to have Cameron; he was an excellent pilot, and I had the utmost confidence in him right from the start. I might as well say that the reason I survived the war was due to having such a good pilot. We did two operational tours together, and not once could he be faulted in flying technique. I was so lucky, and I know, because for various reasons I had to fly with many other pilots, and was soon able to judge the good, the ordinary and the bad. I had my first flight with Cameron on 18th November, 1941, and first operational night patrol with him on 22nd November.

In late 1941 and early 1942, there was little enemy air activity over the east coast of England; most of the action was further south. Therefore a detachment of No. 255 Squadron was attached to No. 29 Squadron at RAF West Malling in Kent, to help out in the defence of London. From West Malling, the first Beaufighter victory was scored for 255 Squadron when on a dark night of 14th January, 1942, Cameron and I shot down a Dornier 217 into the sea about thirty miles off Ostend. We each received a 'flimsy' from the Air Officer Commanding (AOC) No. 11 Group, which read as follows: 'Hearty congratulations on your successful battle last night, well done, Leigh-Mallory.' By now the squadron had destroyed ten enemy aircraft at night, confirmed, and had claimed three probably destroyed, and eight damaged.

Soon after this we returned to Coltishall. On 12th February the *Scharnborst* and two other German warships made their successful escape dash from Brest. We did not take part in the actual attack on these ships, but afterwards spent many hours flying over the North Sea searching for dinghies of shot-down aircrew. Unfortunately none were found.

In the UK a night fighter squadron was usually divided into two flights: A and B, and each flight did forty-eight hours on duty and then had forty-eight hours off. The flight on duty did not leave the station, and as well as being always on call, spent the hours between dusk and first light down at dispersal near the aeroplanes. Therefore, if they were not already in the air on patrol, they were ready to take off at a moment's notice, known as a 'scramble'. The patrols were generally of three hours duration. In winter the duty flight could be as much as sixteen hours at dispersal, in a wooden hut with a few beds and some chairs, and with only a short break for a meal during the night. On days of thick cloud, crews were sometimes called to immediate readiness because of the better chance of following an enemy daylight raider in cloud with airborne radar, rather then using day fighters who were without this apparatus.

When on readiness down at dispersal we were either sleeping (or should I say dozing?), reading or playing cards. There was always a card school going, usually poker, and pennies were won or lost. It needed a strong young constitution to get by because of eating and sleeping at unusual times, and still be fit enough to do one's job when next in the air. However, the amount of leave was adequate, and aircrew were housed either in the officers' or sergeants' mess, where, in both, they were well looked after. They were even issued with extra rations of chocolate and dried fruit, and in our case there was a night flying supper for the flight on duty, which usually included bacon and egg. I remind you that this was wartime Britain, where there was strict rationing of food for the civilian population.

However, even the officers' and sergeants' quarters on RAF stations could vary. I remember at Hibaldstow, my first station, it was just draughty huts to live in, and it was like entering paradise when the squadron moved to Coltishall, which was a permanent station with brick walls, built in 1939. The sergeants' mess there was excellent, the seniors were warrant officers of long service, and they knew what to expect in comfort, and what they were entitled to, and therefore we all gained from this.

A lot has been said about the supposed bad feeling brought about because of the long service regular NCOs in the RAF having to share the same quarters with jumped-up wartime-only sergeant aircrew with very little service. But I certainly did not experience this at all. It is possible that in Fighter Command this bad feeling did not exist because of the small numbers involved, whereas in Bomber Command, with thousands of aircrew on each station, surely the conditions were different, and there might have been trouble.

Anyway, I was enjoying life to the full. I had been very keen on flying and serving as RAF aircrew since at least 1938, and here I was in 1941, not of course as a fighter pilot as I originally wanted, but as the next best thing, as crew, with a flying brevet and sergeant's stripes up and on operations. It was a fine life as only those who have served on an RAF squadron would know.

Social life outside the mess when stationed at Coltishall meant visits to Norwich. As well as the pubs, there was the theatre, and cinemas, and the Samson and Hercules ballroom. Transport into Norwich from the station was provided, usually a three-ton lorry with minimal seating, and in which, if standing, we were flung about at the mercy of the driver, who knew well enough what he was doing! After a few beers and a dance the problem was getting back to camp again, because after taking a girl home it could be very late, and the last transport back to camp was at 11p.m.

In fact I spent many a night on a settee thoughtfully provided in the lounge of the Bell Hotel (or was it the Castle?) and being awakened next morning by the noise of the Hoover going to work on the carpet. Eventually, I found the best thing to do was to make my way to the printing works of the *Eastern Daily News* in Norwich. Here they had an air raid shelter equipped with bunks, and one was allowed to sleep until 4 a.m. when one was awakened with a cup of tea. Then one was driven to Scottow crossroads in the van making the early run to deliver the newspaper to the Norfolk coastal towns, and from the crossroads it was just a mile to the camp gates.

In March 1942, No. 255 Squadron left the joys of being stationed at Coltishall, and was sent to High Ercall, in Shropshire, across the country. This was a wartime-built airfield, and was not at all comfortable in respect of living conditions. Not only this, but the deep mud was a distinct hazard if by chance your aeroplane came off the concrete runway. As well, there was a nasty 900-foot hill, called the Wrekin, not far from the end of the runway.

Another thing I remember about High Ercall is that I was 'put on charge' there! One night Cameron and I were scrambled, but there was some fault when the engines were started up. Directed by the ground crew we dashed off to another aeroplane, and were hardly airborne when the back hatch blew open. Lying on the floor was a parachute left by the previous occupant. It fell out of the open hatch. There was an enquiry, and I was found guilty of not clearing the aeroplane before take-off, which I thought because of the circumstances was a little unfair. I was given the sentence of a 'reprimand', and the previous occupant was given a 'severe reprimand' for leaving his parachute in the aeroplane. I can only hope the parachute did not fall on the head of anybody on the ground, and that the material augmented the clothes ration of whoever found it.

I think there were two reasons for No. 255 Squadron being moved from Coltishall to High Ercall. The first was that, in spite of some success at night against the Luftwaffe, the squadron was in disgrace because of the number of flying accidents that had occurred while at the former. These included two squadron and one flight commander killed in the short time we were there. This was too much for Fighter Command. The second reason was that No. 68 Squadron, stationed at High Ercall, was led by Wing Commander Max Aitken, the son of Lord Beaverbrook, who had some influence at Fighter Command. He, no doubt, had been pressing for a more active role for his squadron in the front line, which was more likely to be on the east coast of England. Therefore the two squadrons were swapped over.

Much of the blame for the flying accidents on No. 255 Squadron must have been due to the fact that the airframe of the Beaufighter II aeroplane did not match the Rolls-Royce Merlin in-line engine. For example, one of the habits of this aeroplane was to swerve violently to the left on take-off, which took some getting used to. We were very relieved when, soon after we arrived at High Ercall, the squadron converted to the Beaufighter VI aeroplane with Bristol radial engine. Both these aero-planes had forward-firing guns, usually four cannon and six machine-guns. I believe that some of the Coastal Command Beaufighters also sported a single machine-gun facing back out of the navigator's bubble on top of the aeroplane.

Incidentally, in the days of 1941 to early 1942, the aircrew of an RAF squadron still took an active part in the burial procedure of squadron members. The officers marched by the side of the gun-carriage on the way to the cemetery, and the sergeants handled the coffin, including the actual burial in the grave. It was not very nice to be used in this way to bury ones comrades, and eventually the RAF woke up to this fact. The procedure was then changed to voluntary attendance, and a burial party from the station did the honours. I suppose the RAF regiment, after a time, took over these duties.

No. 255 Squadron's three months at High Ercall could be described as 'desultory operations', because there was not much enemy activity at night in the area at that time. However, under our new commanding officer, Wing Commander D.P. Kelly, much practice flying took place, both by day and by night. Doubtless his instructions were to lick the squadron into shape. Cameron and I were flying an average of ten night patrols, each of two to three hours duration, plus twenty day flights per month, with many practice radar interceptions included.

In July 1942, the squadron was moved to RAF Honiley to try to protect the Midland towns. The night patrols were stepped up even more, and four enemy aircraft were claimed as damaged by the squadron.

Birmingham was near enough to be visited on our evenings off duty, and I remember going to the Assembly Rooms, where Teddy Foster and his band played for dancing. I tell you there was no lack of female partners, but getting back to camp late at night, as usual, was difficult.

On 19th September, 1942, No. 255 Squadron ceased to keep an operational state, and we went into training for overseas service. The first shock was the issue of khaki battledress, but we kept our 'best blue' uniforms. Soon all hands were engaged in stripping and rubbing down the aeroplanes to be changed from black paint to day camouflage for the journey to wherever we were going. All the air and ground crews were involved in toughening-up exercises, route marches, and ground firing on the range (the aircrew were issued with Smith & Wesson .38 revolvers). Some took the army battle course. The aircrews were trained to re-arm and refuel the aeroplanes, and we navigators were given a crash refresher course in air navigation.

Chapter IV

'There are three rules for writing, and nobody knows what they are.'

(Somerset Maugham)

On 13th November, 1942, eighteen Beaufighters of No. 255 Squadron took off from Honiley and flew to Portreath in Cornwall, waved off by the remaining aircrew, the ground crew and the Honiley station staff. It is thought that some members of the squadron were relieved to get away from Honiley without being in trouble, or shall we say not being found out. As you can imagine once it was obvious we were going overseas no local girl was safe, and the drinking and the partying in the local pubs was without precedent.

At Portreath, loads of maps and charts were handed out to the bewildered aircrew. After a year of night fighter operations in the confines of the UK, and being controlled in our movements by radar, it was all rather strange, and the hope was that somebody knew the way to Gibraltar. I joke, of course!

Happily, all the crews of 255 Squadron made it to the Rock, with perfect flying weather all the way. This part of the journey took five hours and some pilots, worried about fuel consumption, illegally cut across neutral south-west Spain. On arrival there we found we were behind squadrons of Wellington and Bisley bombers waiting to land, and going round and round the Rock at various heights for two hours, without entering Spanish air space, was quite difficult.

The runway at Gibraltar in those days was very short, and the airfield was packed with aircraft wing tip to wing tip, they were parked to within a few feet of the edge of the runway. When we eventually landed there was no room for manoeuvre, and the moment we stopped our engines, the aeroplane was pushed into its allotted place by a team of airmen. We were provided with a meal and a bed for the night, but nobody had told us that only sea water would be available in the showers, and there was no salt-water soap available. However, this minor difficulty was soon forgotten because of what was to come with ablution problems.

The next day the squadron, suitably refuelled, flew on to Algiers Maison Blanche airfield, ostensibly escorting DC3 aeroplanes containing American paratroops who were to be dropped at Oran. At Maison Blanche, only a few days after the ground troops had landed on the coast of North Africa, chaos reigned. Perhaps the toughening-up exercises undertaken so unwillingly at Honiley had not been such a waste of time after all. There was nowhere to sleep, so we had to make do with a single blanket to cover us, on the concrete floor of the airport building, or on the grass underneath our aeroplanes. We queued for ages at the airport cafe, where one harassed RAF cook with the assistance of M. le Patron, tried to keep all supplied with food. Water for washing and shaving was collected in cut-down petrol tins from a cold water tap outside the airport building. I am happy to report that only after a day or so of this we were billeted in some huts vacated by the French.

If nothing else 255 Squadron aircrew will never forget having to refuel their Beaufighters from four-gallon petrol tins. To fill each aeroplane required some 500 gallons of petrol! I opened each tin and handed it up to Cameron on the wing, who poured it into the tanks, through a filtered funnel. It was exhausting work in hot sunshine dressed in winter-weight khaki, and seemingly never ending. None of the aircrews had anticipated life without petrol bowsers and ground crews to operate them.

We were lucky enough to have some help with minor repairs to our aeroplanes from the ground crews of the Spitfire squadrons with whom we shared the airfield. Incidentally, our own ground crews were still living it up at Honiley. It was not until 26th November that they entrained for Liverpool and went aboard SS *Maloja*. On 7th December this ship arrived at Algiers, and was then sent along the coast to Bone port where they disembarked.

For some reason, someone in authority had decided that we should proceed to North Africa without our airborne radar sets; they were removed from our aeroplanes at Honiley just before we took off. Not only this, but there was no GCI ground control when we arrived at Maison Blanche; therefore, for the first vital ten days of the landings we were as good as blinded at night. We flew patrols, of course, because there were continuous German and Italian bombing raids on Algiers docks and Maison Blanche airfield, but with no successful interceptions by us. Standing patrols at night without ground control and AI sets in the aeroplanes were quite futile. As our Mark IV AI had been in use for years, and was no secret to the Germans, it must have been a complete idiot who determined that the radar sets should be removed from our aeroplanes. Instead we would be followed out by the ground crews by sea. After all, Mark IV had been standard equipment for some time in Malta, and at the eastern end of North Africa.

Regarding the above, Roger Kirkman (a Sergeant ground crew radar mechanic) puts forward the valid theory that some people in authority were very worried that the capture of a complete AI set would facilitate the development of jamming equipment by the Germans. However, as he says, the obvious solution was to fly out the sets in a Fortress aeroplane (to the squadron), as indeed happened on 27th November, but this was much too late.

Without radar there was very little warning of the approach of enemy aircraft at night, and although there were brave attempts by 255 Squadron pilots to get the Beaufighters, not already on patrol, off the ground at the last moment to avoid the bombing of the airfield, nine were a write-off, and others were damaged during the nightly raids. It was during one of these raids that a 255 Squadron pilot, Hugh Wyrill, was killed as he was climbing into his aeroplane. His navigator was badly injured.

On 22nd November a ground controller arrived together with the makings of radar ground control. Shortly afterwards a small advance party of our ground crews turned up, but still there were no airborne radar sets. By this time, because of the bombing, the squadron was very short of serviceable Beaufighters. The remaining three or four were flogged to death on night patrols, but still no interceptions were made. As you can imagine, we were completely frustrated by this: there were all these enemy aeroplanes flying about overhead, doing a terrible amount of damage, and causing many casualties. We had been sent out to shoot them down and we could do nothing about it.

On 24th November, No. 89 Squadron, a very experienced

night fighter unit stationed at Abu Sueir in Egypt, received Emergency Signal A304. This was an instruction that six of the Beaufighters and aircrews, and ten ground crew, were to proceed to Maison Blanche airfield via Malta, to take over from us the defence of Algiers. This squadron had been equipped with Mark IV AI for a long time, so why were we not allowed to bring our radar sets in our aeroplanes with us? Between 25th November and 12th December, 1942, the 89 Squadron detachment shot down nineteen Heinkels, Junkers and Dorniers in the defence of Algiers, and if it had not been for the clot who ordered that we should not carry AI on the journey out, these kills would have been ours.

As mentioned above, on 27th November a Flying Fortress arrived with our radar sets as part of its cargo, but by this time we had very few aeroplanes left in which to install them. Therefore, it was decided that twelve crews should be flown back to the UK to collect new aeroplanes. We travelled in a DC3 to Gibraltar and thence to Portreath in a Flying Fortress, both American crewed. The latter flight was quite an experience as it involved flying at a very low level over the Bay of Biscay to avoid the Luftwaffe patrols out from western France. In less than one week we had been issued with new Beaufighters at Filton, near Bristol, and after the usual tests to make sure everything was working properly, we set out again for Gibraltar. Incidentally, the short cut across south-west Spain on the original trip had caused a diplomatic furore, and when we went through Portreath the second time we kept our own counsel when the intelligence officer there implored us not to act like some idiots had a short time previously.

On arrival at Gibraltar the queue for landing was shorter than previously. The undercarriage on Val Phillips' and

Dennis Pollard's aeroplane refused to lower. After an impeccable wheels-up landing on the then short Gibraltar runway, Val ended up a few feet from the end. To their astonishment, after being told to immediately leave the aeroplane with their kit, a crane appeared and tipped their Beaufighter into the sea. Packed out with aeroplanes for the North African battles, at that stage there was no room on the Rock for any that were unserviceable.

At Maison Blanche, it was with great satisfaction that the squadron took to the air on patrol now fully equipped, but by this time the enemy had reduced operations against Algiers, and was bombing Bone and other places further east.

Another strange strategic decision that affected us was when Royal Air Force Command in North Africa was not provided with any personnel transport aeroplanes. As the Beaufighter was the only aeroplane in the area that could carry passengers, albeit uncomfortably, the squadron found itself ferrying passengers about, sometimes on very dangerous missions, when we should have been resting in anticipation of the next night's operations.

An example of these extra duties is that on 21st November, the air officer commanding, Air Marshal Sir W.L. Welsh, needed to fly from Maison Blanche to Bone, the Allied occupied airfield nearest to the front line. Therefore, he was flown there by our CO, and Cameron and I were deputised to take along his staff officer, Group Captain Long; the escort was six Spitfires.

During the one hour and twenty minute journey, somebody was heard to disclose over the open R/T (radio telegraph) that he had seen two Beaufighters with Spitfire escort flying east. We landed at Bone, where we were instructed by the airfield CO that in the event of an attack we must get the Beaufighters off the ground immediately. Sure enough, after about half-an-hour, the airfield klaxon sounded an air raid warning, and almost at once Messerschmitt 109 fighter-bombers came in at low level to attack.

They went for the Spitfires first and soon many were burning around us. The ground crews were in the slit trenches and who could blame them: so there was nobody to help us get off the ground, but the pre-overseas training I had received had taught me how to prime the engines of a Beaufighter in an emergency, and they started immediately as they were still warm, and I then hastily clambered up into the rear hatch.

Both Beaufighters took off, but Wing Commander Kelly's aeroplane was attacked and a hydraulic pipe was shot away. With the fluid spraying all over the place he thought it inadvisable to land again at Bone, and therefore flew back and crash-landed at Blida airfield, near to Algiers.

Our take-off was most dramatic. As I climbed into the aeroplane Cameron started rolling, but we were not on the runway, and to clear a burning Spitfire and some trees he put down about fifteen degrees of flap. In his hurry to get off the ground he had not checked his seat lock, neither was he strapped in. His seat collapsed and the aeroplane commenced to climb steeply and he had no control. Hearing the agonized shouts from the front I rushed forward and somehow was able to lift his seat into the upright position, and he was then able to regain control and avoid the fatal stall.

I went back to my seat and the first thing I saw was an Me 109 on our tail. I shouted a warning to Cameron and he, knowing he could not out-turn an Me, flew down as low as possible and headed inland, forcing the Me to increase the distance from its own base. After some extremely dangerous evasive action at this low level, the Me gave up and broke away, probably worried about his fuel supply. We contacted Bone airfield and were ordered to land again, where we collected both the air marshal and the group captain, and flew back to Maison Blanche.

On 5th December, 1942, all available 255 Squadron Beaufighters were flown forward to the most advanced Allied airfield in Tunisia, which was near the village of Souk-el-Arba (Thursday market). We were here to defend the Port of Bone, where the ground controller Squadron Leader Brown, OBE, had set up his GCI station. At Souk, which was a metal mesh runway laid in quick time by the Americans, we were billeted in the Arab village, and massive amounts of A1 63 powder was needed to protect us from the bugs. Freddie Lammer, our navigator leader, claimed that, armed with a torch and a piece of wood, he destroyed 200 in a one-hour session.

On the night of 5th December Geoff Humes, with navigator Johnny Sayer, opened the score for 255 Squadron in North Africa, by shooting down two Heinkel 111s, and during that December a further sixteen German and Italian aircraft were destroyed at night. Cameron and I shot down a Heinkel 111 on the 6th, and for the record here is the report according to the Public Record Office at Kew:

Pilot F/O J.C. Cox Navigator F/Sgt P.L. Croft Take-off 20.55 hours. Flew at 14,000 feet towards Bone on vector 300. Patrolled along 280 and 100 until 22.30 hours. Various vectors were given after a bandit. Radar contact was made after 10 minutes. Several corrections were made, which included a 1000 ft.

climb, and then a visual was obtained at a range of 2500 ft. on exhausts. Beaufighter closed in and E/A was identified as a Me111. Opened fire at 150 yards range with one short burst (cannon and machine gun). The E/A caught fire in the fuselage and port engine. Closing in to 50 yards another short burst was given when the E/A became a ball of flame and blew up. Evasive action had to be taken to avoid debris. E/A was seen to descend vertically and hit what was thought to be the sea. Contact had been made at maximum range, and when Beaufighter finally closed in E/A was doing only 170/180 mph. Some of the debris landed on the Beaufighters [*sic*] wing where it burned until blown off. Claim 1 Heinkel 111 destroyed, confirmed by pilot (S/Ldr Player) and R/O of another aircraft nearby.

On 8th December we were most pleased to welcome the main party of our ground crews who had come ashore from SS *Maloja* at Bone Harbour. Especially, it was a great relief to have the aeroplanes properly serviced at last. Soon a tented camp had been erected, and most importantly, the latrine arrangements were put in order with several home-made wooden six-seaters available. Heating water for shaving and other purposes, in cut-down petrol tins, filled with sand soaked in petrol and then fired, was in vogue, but unfortunately this caused some minor casualties, including Kevin O'Sullivan losing his hair, and Arthur Woolley destroying his, and others' kit, by setting fire to their tent.

Now we had our own cook and his helpers there were no complaints about the food. Basically it was 'compo' rations, i.e. large boxes containing all that was necessary for fourteen men for one day, or one man for fourteen days. Some of the tinned puddings were delicious. These rations were augmented by local products, particularly meat, eggs and oranges. Therefore, we might have lived in a reasonable state of comfort except for the weather. Talk about sunny North Africa! It was day after day of high winds and rain; sometimes the airfield was a sea of mud, and we took to operating only in emergencies.

As well as the weather, it was also most unpleasant at Souk because the Germans had spotted the airfield and came over from Sicily to bomb and strafe during the day. Unfortunately we lost a ground crew airman, LAC Owen, killed in one of these raids. Also two Beaufighters were destroyed on the ground, and several were damaged. After this even the previous most nonchalant aircrews of 255 were seen to be frantically helping with the digging of safety trenches!

In the meantime it had been decided that the main headquarters for the squadron should be at Sétif, a French Air Force base in Algeria, but a detachment would be maintained at Souk-el-Arba for the continued defence of Bone, where much equipment for the British and American armies was being unloaded. At Sétif, the ground had been properly drained, but it was higher up, and in February it snowed there. It was then decided perhaps to give up Souk altogether, and therefore Cameron and I, and a small party of airmen, were sent to Duzzeville, south of Bone, to look into the prospects of flying from a new Summerfield track landing strip to be called 'Tingley', so named after the army major who was responsible for its construction.

Duzzeville was in the centre of the wine-producing area of Algeria, and the detachment was billeted on top of wine storage vats. Apparently the inside of the concrete vats were coated with a special substance to prevent the wine soaking away. Needless to say, within a very short space of time somebody produced a length of piping long enough to reach the new red wine through the unlocked hatches of the vats, and a good deal was consumed by the detachment.

Others of the squadron joined us, and we were there over Christmas 1942. We thought we did rather well when we sat down to a chicken Christmas dinner, the birds purchased locally by Squadron Leader Peter Dunning-White, the senior officer. It was not until we rejoined the squadron at Sétif early in January 1943 that we learned they had fared even better with a choice of turkey or steak and kidney pie.

There were further successful battles by the squadron crews in January and February, 1943, but in spite of many operational patrols by Cameron and me, nothing came our way. We seemed always to be vectored on to our own Wellington bombers returning from raids on Italy and the Balkans to their bases in Algeria. After the Germans were cleared from north Africa the squadron moved over into Sicily and then into Italy and continued to do well. Eight Distinguished Flying Crosses (DFC) were awarded to members of the squadron, but unfortunately crews were lost, shot down or in flying accidents.

Overseas, No. 255 Squadron destroyed eighty-five enemy aircraft (confirmed), with claims for four probably destroyed, and eleven damaged. If added to those in the defence of the UK, the final score for the squadron during World War II was ninety-six enemy aircraft destroyed, with claims of seven probably destroyed and thirty-one damaged.

Chapter V

'He has not always been able to judge between what he thought had happened, or wished had happened, and what actually occurred.'

(Lord Stewart of Fulham describing the Richard Crossman diaries)

It was in March 1943 that I really began to feel the effects of flying operationally. I had by then completed over eighteen months of operations, which was I believe the limit for night fighter crews, as was the equivalent thirty operational trips for bomber crews, and 200 hours operational flying for day fighter pilots. I suppose if the squadron had continued in a defence role on operations for which it had been trained the fear factor might never have surfaced, but we had heard that other operations, such as intruding over enemy territory, were being contemplated for the squadron by the Powers That Be.

Other things prompted me to request a rest: one was that I was having to resist a strong inclination to throw myself out of the aeroplane on moonlit nights over the sea, so much so that I flew all the time with my parachute strapped on. Another was that I thought I was having trouble with my eyes. I asked the squadron doctor Flying Officer Brennan for help with the latter (I thought he would not believe me with the former!). I was sent to

39

the hospital at Constantine and was examined by the eye specialist, who could find nothing wrong.

After I returned from Constantine I applied to see the squadron commander, now Wing Commander Johnny Player, and he was most sympathetic about my problem. It was arranged that I should come off operations, and indeed I was to be sent back to the UK with other tour-expired aircrew of the squadron. Cameron, of course, was not at all pleased about his navigator disappearing, but continued to fly with one of the spare navigators on the new role for the squadron which mainly was intruding by night over Sicily, looking for ground and air targets. According to the squadron records the results were disappointing, and several 255 crews were lost on these operations. As well, the squadron took part in daylight operations protecting convoys of ships, and doing other patrols in daylight over the Mediterranean, looking for German Junkers 52 troop transports to shoot down. All this looked jolly dangerous to me, and I was glad to be out of it, and awaiting posting home.

In May 1943 I was sent back to Algiers, and eventually sailed in the good ship *Franconia*, which, with other ships in convoy, and an escort, was bound for Liverpool. Considering we were sailing on a troopship, we had a most comfortable voyage, with four sergeants sharing a cabin. We had plenty of space on the upper decks whereas down below were about 3,000 Italian prisoners of war, crammed into the minimum amount of space with scant ablution facilities. Obviously there was a limit to the numbers to be carried on a troopship by law, but where they were housed was another matter. We sergeants had certain policing duties to carry out, and I can tell you, the stench coming from those poor devils in their small space was so strong that the time we spent on patrol in the lower decks was reduced to a minimum. The convoy took as long as eight days to sail from Gibraltar to Liverpool because we went far out to sea in the West Atlantic. This was to avoid German air attacks from their bases in western France.

My experience of RAF records during my service was always first class; others may have a different view. When our ship docked in Liverpool we were sent to a transit camp, where, incidentally, I heard reveille sounded on the tannoy at six o'clock in the morning, the only time in nearly six years of service! Anyway, the next day it was quickly established who I was, and what I was to do. After ten days leave I was to report to No. 62 Operational Training Unit (OTU) at Ouston, near Newcastle, as an instructor.

When I arrived at Ouston, I found I was to be employed teaching would-be navigators the mysteries of air interception by radar actually in the air. In the three months I was there I flew fifty times, and completed 142 hours flying in the Anson aeroplane, each time with two pupils.

Socially, I had a super time there, often being entertained in the WAAF sergeants mess – quite illegal of course – and I had a lovely WAAF girlfriend to take to Newcastle and the local pubs. It was great to get back to English beer, even the wartime brew, because the NAAFI issue of French beer in North Africa had always tasted slightly of onions. Incidentally, when we first arrived in Algiers, the price was one franc per bottle; when the NAAFI arrived and took over the brewery, it immediately advanced to six francs!

* * *

Unfortunately, the bliss of serving at Ouston did not last long before Cameron, back from North Africa, requested my presence at No. 63 OTU at Honiley, where he was instructing pilots. He had suggested that we crew up together for our second tour of operations, and as I was most anxious to continue to have him as my pilot, of course I agreed.

After my experiences in North Africa I was keen to be commissioned, particularly in case of a further posting overseas, wherever it was. Because I was well known to the senior instructing staff at No. 63 OTU it should have been easier and quicker to be granted a commission there than at Ouston. Therefore I applied for a posting to Honiley, which was allowed. Also, Cameron wished me to be in the same mess as him for our second tour, which was sensible reasoning.

I must say that when I arrived at Honiley and found that the training aeroplanes were Beaufighter IIs, with which we'd had so much trouble two years previously on No. 255 Squadron, I was rather put off, but as the instruction I was to give in the air would be in daylight, the risks would be much less. No. 63 OTU was an advanced training unit, and although I gave AI demonstrations to the navigator pupils, and also tested some for their pass marks in the air, other times I was just crew because the pilots were being instructed or tested up front.

In the four months I was at Honiley – from September 1943 until January 1944 – I flew just forty-six times, and completed only forty hours of flying. It does not sound much, but there were also many hours of instructing on the ground to take into account.

The most important reference for a commission when

already serving as aircrew in the Royal Air Force in wartime is the first, i.e. the officer who knows you. After this, as the application proceeds upwards, the others involved usually sign because they know the first officer, and not necessarily because they know the applicant.

At No. 63 OTU the officer commanding No. 1 Training Squadron, in which I was an instructor, was Squadron Leader Johnny Wright, with whom I had crewed as a navigator in the early days on No. 255 Squadron, and he, indeed, had been a senior officer with the squadron throughout my first tour.

I filled in the application papers, and from then on it was handled by the Orderly Room. After Johnny Wright had approved me, the wing commander of the OC Training Wing, and then the station commander also signed, and then the application moved on to Group. After a little while I was summoned to Group Headquarters, and in my best blue, with everything brushed and polished, I appeared before an air vice marshal. The AVM asked me about my service so far, and then gave me a lecture on the extra responsibilities facing me if I became an officer.

On 11th December, 1943, the Orderly Room informed me I was now commissioned as Pilot Officer P.L. Croft, No. 170341. I took a few days leave to purchase my new uniform, and on return to the station packed my bags and took them from the sergeants' mess over to the officers' mess. Yes this was all that was necessary for newly commissioned aircrew in wartime. There was no time to go to an officers training unit, I just moved over to my new mess, to be greeted with some ribaldry by my friends already installed there.

*   *   *

Early in 1944 Cameron determined it was time to go back to operational flying, so we discussed what we should do. Before we could take any steps towards this, we received individual notices of posting to No. 140 Wing, this was part of No. 2 Group, flying Mosquito aeroplanes on daylight missions. Apparently a previous flight commander on No. 255 Squadron, Peter Dunning-White, now wing commander in charge of flying personnel at No. 2 Group, had been searching for experienced aircrew for the group, and had put our names forward.

When we arrived we found a collection of aircrew, each of whom had a minimum decoration of a single DFC. It was obviously a very talented band of flyers. Listening to these chaps in the mess, Cameron and I became more and more troubled. We were completely untrained for this type of flying, and how long would it take to learn on a squadron, as part of a wing, and not at an OTU?

It is not a good thing for one's career in the RAF to turn down a posting, particularly if it has been initiated at Group, and it took a little while for Cameron to make up his mind, because as early as this he was thinking in terms of making the RAF his permanent career after this war was over. But, eventually, to my great relief, he requested an interview with the commanding officer, Group Captain P.C. Pickard, DSO, DFC. The group captain was a veteran of Bomber Command, with several operational tours already completed. He was also, incidentally, the star of the government propaganda film about the command, titled *F for Freddy*.

One morning we were shepherded into the great man's office, and were welcomed to the unit. Then Cameron had to confess that the posting was a mistake, that we were a night fighter crew, and it was our intention to apply for another tour of duty of this type of operation. Without further ado the group captain called across to his adjutant, 'Post these two officers to where they want to go,' and the interview was over.

Unfortunately, I have to report that Group Captain Pickard did not live long after this. He led the famous raid on Amiens prison in which 120 Frenchmen were incarcerated, condemned to death by the Gestapo for helping the Allies. The Mosquitos knocked down the walls of the prison by superb precision bombing, and some prisoners did escape, but Group Captain Pickard crashed, and he and his navigator, Flight Lieutenant Broadley, DSO, DFC, DFM, were both killed.

Chapter VI

'A writer should never write about the extraordinary. That is for the journalist.'

(James Joyce)

When Cameron and I had completed our first tour on No. 255 night fighter squadron, we had returned from Tunisia to the UK tour expired and we were posted eventually to No. 63 Operational Training Unit as instructors. After about six months at the O.T.U. Cameron decided he wanted to join an operational squadron again. As a fully paid up member of the Navigators Union I was not enthusiastic, but because I did not want to lose Cameron, who was a first class pilot, I agreed and after our experience at No. 140 wing we thought we should find a night fighter squadron to join as soon as possible.

We then found out that Johnny Wright was a flight Commander on No. 68 squadron, and not only this but the commanding officer of No. 68 was Wing Commander D. Hayley Bell, a former flight commander on No. 255 squadron. It seemed that No. 68 squadron was just the place that we might feel at home. Therefore Cameron telephoned Johnny Wright and enquired as to whether they required another crew. They did, and in due course Cameron and I received our official postings. Talk about the old-boy network, it certainly worked in the night fighter world of the RAF in the Second World War.

Anyway, early in 1944 Cameron and I joined No. 68 squadron for our second tour as a night fighter crew.

When we arrived on 68 we found that half the squadron were Czechoslovaks, in fact the great majority of 'A' Flight. Like everybody else we had heard about the gallant Polish airmen, and many others, who had joined the British forces to fight the Germans, but few had mentioned the Czechs, therefore it was something of a surprise to find them well established on 68 Squadron. Even Czech Major-General Vlad Nedved in his very long article published in 1999, in *Intercom*, devoted only one line to the presence of Czech aircrews on night fighter operations. However, I am told it was the Free Czech Government that required that every branch of the Forces should include Czechs, and in 1940 No. 68 Squadron was being formed, and therefore Mansfeld, Klouchnik, Chabera and others were posted in to make a Czech flight.

Miro Mansfeld was a super chap with a great sense of humour, and I got to know him quite well, possibly because his English, both spoken and written was much better than that of most Czechs on the squadron. His favourite word was 'blooody'. You will notice the three 'o's, which is the nearest I can get to his pronunciation of the word. I shall always remember him saying – when the others had described how they had got away from Czechoslovakia by stealing an aeroplane or by riding under a train how he 'bloody well walked'. Also, when describing life at home with Bobina, his wife. 'Here am I, Squadron Leader Mansfeld DSO, DFC, AFC, a great hero, and who has to make the coffee at home, I bloody do.'

Miro shot down many German intruders during his two tours on 68 Squadron, as well as VI flying bombs and shot up the odd E boat, but his shooting down of a V1 on the night of 29th July 1944 was a classic. He took off at 01.40 hours, and commenced to climb over Castle Camps as instructed by Ground Control, when he saw a V1 lit up by searchlights. At 01.45 he shot it down, cheered on by 'A' Flight ground crews watching from dispersal.

I do not think that Miro and his radar navigator Slavek Janacek, held the best record for shooting down Germans. After all, WO (commissioned later) Bobek, with Sergeant Kovarik, destroyed four Do217s and one JU88 and claimed one probable and three damaged during the month of July 1942. The Czechs were very good fliers and good shots.

I have a copy of a telegram sent to us around October 1969:

To the co-fighters of the 68th Squadron
Dear friends,
At the occasion of the 25th anniversary of the final victorious flights of the 68th Squadron we are sending to you our best wishes to spend an agreeable joint evening. We in Czechoslovakia shall be in memories with you. Simultaneously when you raise your glasses to a joint toast we would like to drink in Czechoslovakia with our Pilzner beer to the health and success of all members of our Squadron. We are convinced that some time in the future 'Reunions' we shall be able to participate personally.
Yours. Co-fighters in Czechoslovakia.
Ladislav Bobek, Josef Vopalecky, Frantisek Chabera, Jan Serhant, Josef Adam, Miroslav Jiroudek, Karel Seda, Karel Bednarik, Ferdinand Gemrod, Vladimir Cupak, Badrich Kruzik

As previously mentioned, we had reported to No. 68 squadron, then flying from RAF Castle Camps near Lincoln. It was one dark, late afternoon when we arrived and, after being shown to our quarters and unpacking, we went over to the mess and the bar. As we stood rather hesitantly in the doorway, a slim, rather baldish chap came over to us and said, 'You must be Cox and Croft; I am Leslie Kohler the I.O. of the squadron.' It was 'Koke' Kohler and he made us most welcome and introduced us to the early evening drinkers at the bar.

Incidentally, Koke and I became good friends, which continued on into peacetime. When I left Gravesend in 1958 I shared a flat with him (by then a widower) and his two sons in Richmond for two years, until I moved into my own place in Twickenham. He was my best man when I married in 1968.

Cameron and I spent fourteen months serving on No. 68 squadron, but although we did a lot of night operational flying, no enemy aircraft came our way. We were in B flight, which in the main was English; practically the whole of A flight was Czech. No. 68 was another of those night fighter squadrons formed so hastily in the winter of 1940–1941, and the Czechs who were greatly experienced and resourceful flyers gave such character to the Royal Air Force. Many of them had escaped from the Germans and had then joined the French Air Force. At the fall of France they had finally joined the RAF.

By this time the No. 68 Squadron score stood at forty-six enemy aircraft confirmed destroyed, with claims of ten probably destroyed, and twenty damaged, as well as three E-boats. The shooting down of twenty-four of those aircraft destroyed took place between March and August 1942, when the squadron was stationed at RAF Coltishall, under the control of the ground station Neatishead, as well as Happisburgh CHL (Chain Home Low). The Luftwaffe at that time was passing over the east coast of England on its way to try to bomb the Midlands towns; it was also actively bombing the east coast towns.

During this time the squadron had been very successful, particularly on the night of 23rd July, when it shot down five enemy aircraft in one night, and then went on to an unprecedented period of operations in August, when a further seven kills were confirmed, and it claimed five probably destroyed, and four damaged.

Unfortunately, when the Luftwaffe made Norwich its main target, due to a breakdown in the defence system, the squadron was not in position, and was taken by surprise. Therefore they were not able to defend its 'hometown' (only ten miles away), and the heart of Norwich was brutally destroyed by about thirty bombers. The 68 Squadron flight on duty was scrambled, but not in time, and the bombers got away unscathed.

Why Norwich? It is said that Hitler, incensed by the fire damage done by RAF bombing to the medieval buildings in Lübeck, the old Hanseatic city in Schleswig-Holstein, ordered that retaliatory terror attacks should be made on certain British cities. These came to be known as the 'Baedeker' raids, so called because the targets were said to have been picked from the tourist guide of that title, and Norwich, as a cathedral city, was one such prime site. However, in the follow-up raid on Norwich, all was ready, and several Dornier 217s were shot down or damaged by No. 68 Squadron.

Coltishall airfield itself was bombed from time to time by enemy lone fighter-bombers coming over in the returning RAF bomber stream, and therefore difficult to identify by our radar screen. The Luftwaffe had commenced using the Messerschmitt 410, which was a great deal faster than the Beaufighter; in fact four Mosquitos of No. 151 Squadron were detached to Coltishall to attempt to deal with this problem.

Although when Cameron and I arrived on No. 68 we found that our new squadron was still flying Beaufighters, it had converted from Mark IV to Mark VIII airborne radar. Dear old Mark IV was all very well and had given sterling service to night fighting, but it was useless at low heights when the massive signal received from land or sea returns obliterated the signal of the aircraft being intercepted. Even at 20,000 feet the maximum range did not exceed four miles. By contrast the narrow rotating ten-centimetre beam from Mark VIII, and later its American produced version, the Mark X, gave a maximum range of eight miles at all heights above land or sea, and was a revolutionary advance for the squadron.

At Coleby Grange, in February, the weather turned bitterly cold with deep snow, and a photograph in the squadron records shows Carol Seda, Gilbert Wild, Johnny Wright, Jack Haskell and me, armed with spades, helping to clear the runway. Maintenance of the Beaufighters continued to be excellent in spite of the inclement weather, and one had the greatest respect for the ground crew trying to do their jobs in the open in that sort of climate.

In March 1944, the squadron was moved to RAF Fairwood Common, near Swansea; as some wit remarked, defending the Gower Peninsula! The main role for us there was Channel patrols and acting as convoy escorts. Over the water on convoy escort duty many Beaufighters, including our own, were fired on by the navy when taking over from the day fighters at dusk. Fortunately, the navy were even worse at shooting than at aircraft recognition, and we survived. When one heard the noise, and saw the black bursts of anti-aircraft shells in the air quite close, one realised how brave the crews of Bomber Command must be to have to go through that experience on every raid.

At Fairwood our pilots began to convert from the Beaufighter to the Mosquito XIX, and the navigators from Mark VIII to the American Mark X radar. There were no problems in the radar changeover so far as I was concerned. Regarding the change of aeroplane, I must admit that I missed the comfort feeling of the Beaufighter, which plonked down nice and firmly on the runway on all three wheels during landing, whereas the Mosquito was inclined to teeter on its front wheels before the final put down. I also missed having my own space in the centre of the Beaufighter. However, the Mosquito was a faster and better aeroplane, and eventually one got used to it.

The squadron enjoyed the social life at Fairwood outside the mess, especially at the Langlands Bay Hotel dances, until the American forces moved in to compete for the favours of the local girls, and others. We took solace in the pubs of Swansea: I think our favourite was the famous 'No. 10'. This was all very well, but there was a problem in that all the pubs were shut on Sundays! However, there was an answer even to this because we became associate members of the Gower Country Club, the bar of which was open every day; life membership cost us just half-a-crown. One of the members of the Fairwood Common officers' mess, who lived locally, was Flight Lieutenant Alcock of the famous Atlantic Ocean flyers team of Alcock and Brown.

The squadron scored some victories; in May 1944 while

at Fairwood, when Squadron Leader Mansfield and Flying Officer Janacek shot down two Dornier 217s, Sergeants Peters and Rackham brought down one Heinkel III, and Flying Officers Wild and Baker caught one Junkers 88. It was at Fairwood that my old friend Denis Pollard joined the squadron, together with his pilot Val Phillips, to commence their second tour.

During May the squadron personnel were subjected to a series of lectures, which might herald some important events in the near future. They were 'Security', 'Internment of Prisoners of War', 'Grasshopper Procedure' and 'Escape'. On the 28th, Air Chief Marshal Steele, CB, DSO, visited the station and, *inter alia* presented us with the No. 68 Squadron crest. The crest had first been officially requested by Wing Commander Dottridge, when he was commanding officer, and he had decided on an owl as the centrepiece. He had asked the Czechs for a suitable motto, and after some debate the Boy Scout motto 'Be Prepared' was amended slightly to 'Always Prepared', which in the Czech language is '*Vzdy Pripraven*'.

It was at Fairwood Common that the chant dedicated to the special aircrew rations was first heard: 'Aircrew rations, whose aircrew rations? Aircrew rations for Kohler's big, fat, bouncing, bulging baby.' So went the chorus of this unique 68 Squadron doggerel, with each verse separately telling the story of the extra raisins, eggs, chocolate and sultanas, etc. allowed to aircrews. This song was, of course, directed at Koke much to his discomfort. The squadron intelligence officer, in whose hands was entrusted the issue of these products, was the last person to think of diverting even one currant to his baby son Michael. Embarrassingly for Koke, Michael at that time was the fattest and most bulging, baby imaginable.

On D-Day, 6th June, 1944, the Second Front opened with the landings on the French coast, and there was a great deal of excitement in the squadron when we were advised of this the evening before. However, the next forty-eight hours were a complete anticlimax for us because we were told that the air would be so full of aeroplanes directly connected with the Second Front that all other planes in the UK, including ours, must remain on the ground. However, this did not mean that we could all make off to the pub. Our usual readiness state must be maintained, and therefore the poker games and bridge schools down at dispersal kept going at full swing throughout the two short nights of 5th and 6th June, and then, after this, we resumed normal patrols again.

On 23rd June 1944 the squadron moved again, this time to Castle Camps, near to Saffron Walden in Suffolk. When we arrived there the pilots completed their conversion from the Beaufighter to the Mosquito, and the navigators to the Mark X airborne radar. Much practice flying was carried out by the squadron to complete these conversions.

Soon after we arrived at Castle Camps, I was in the mess one day, minding my own business, when the new commanding officer, Wing Commander George Howden, came in and announced that he had heard from Group that the aircrews of the squadron must have a sports officer, and I just happened to catch his eye. It was the cricket season and I found out that Ashdon Village would be pleased to give us a game, and also that there was a load of cricket gear on the station. I had some difficulty in raising a team because 'A' Flight was mainly of the Czech element. Consequently, it was only 'B' flight who took to the field of play on a good pitch, well looked after by the village. Naturally, I made myself captain and I also took the role of wicket-keeper, until the last couple of matches when I was so desperately short of people that I had to ask Humphrey K. Humphreys if he would play, which he did with the proviso that he would be behind the stumps.

The only recorded result of several matches was on 17th August when the squadron lost by four runs; one or two matches were abandoned because of bad weather developing.

Tennis was played on the Watton Park courts, kept in very good condition by the family's gardeners. Later on in the year, in September, the squadron soccer team, captained by me playing at right back, and with Jock McCulloch at left back, beat the village by ten goals to one.

Castle Camps airfield was situated reasonably close to Newmarket, believed to be the only racecourse allowed to hold meetings throughout the Second World War. The local farmers, all horse-racing men, with unlimited supplies of petrol for their farm vehicles, but unable to use their cars for social reasons because of being known to the police, were happy to pour petrol into the tanks of the cars owned by squadron members, and accompany them on visits to the races. Indeed, they were very generous with their hospitality in many ways. Personally, I had many full English breakfasts – at that time of the war quite unknown to most of the civilian population – in a farmhouse kitchen, at a very early hour, having been invited in on my way back to bed from dispersal.

At Caste Camps the squadron's future in the war was clearly defined because the first V1 flying bomb had detonated on English soil and Fighter Command was attempting to cope with this new situation. So, 68 Squadron was to be in the front line for night operations against the V2, known generally in the forces as 'anti-diver' operations.

We are talking about the V1 pulse jet pilotless aircraft launched by the Germans first from occupied France and then from Holland. Needless to say some wit had soon conjured up funny names for them such as 'buzz bombs' and even 'doodlebugs'. It is said that Hitler himself had put extreme pressure on the mechanics working on the V1s to hurry up their manufacture. In his mind it was retribution for the D-Day landings.

By this time the British people had withstood three years of bombing, with particular ferocity in London (including the blitz from September 1940 until May 1941), and they had shown great courage in the face of this terrible carnage. But with the D-Day landings and the war in other parts of the world going quite well, it might also have seemed as though the threat to their lives was over, and even that the war might soon be over. Imagine their feelings when this new terror began raining down on them from the skies, commencing on the night of 12th–13th June, 1944.

On the ground the first indications of the approach of a V1 was a throbbing sound, which gradually increased to a fiendish staccato crackling, and the ground seemed to be trembling under one's feet. Then there was a silence as the engine cut out, and the bomb was on its way to the ground. Next was the roar of the explosion as the bomb hit the ground or whatever was in its way.

As well as the conversion programme, a lot of operational flying was done, as and when the crews became fully adapted to the new aeroplanes and radar. The weather was very poor but the squadron pressed on. Yet it was not until 2nd July that the first flying bomb was seen by a 68 Squadron crew, and a week later before the first was shot down by Wild and Baker. Another three were shot down by the squadron in July including an absolute cracker by Mansfeld and Janascek. Mansfeld took off at 1.35a.m. on the morning of 2nd July, and was about to gain height over Castle Camps according to instructions from the ground, when he saw a flying bomb lit up by searchlights. At 1.40a.m. he shot it down, cheered on by 'A' Flight ground crews, watching from the ground at dispersal.

Unfortunately, we suffered many casualties, most of which were directly attributed to the laws for intercepting the flying bombs. Because they flew so low (approximately 400 feet) and so fast (390 knots I.A.S.) pilots were instructed to fly on patrol at 6,000 feet and then to dive down on to the target in order to have the necessary speed to catch it up on the straight and level. Inevitably, in the dark, and without electronic altimeters (fitted later), mistakes were made, and it was reckoned that six crews of the squadron were killed using this method of attack.

As well as the normal patrols over the North Sea, the squadron also patrolled off the Dutch islands hoping to catch the flying bombs despatched from Holland early on in their flight.

In spite of the large amount of flying, there was much life going on in the mess at Castle Camps. The officers' mess was at Waltons Park, a magnificent country house set in parkland. The Luddington family were still living in part of the house. Needless to say, the family heirlooms had been removed from the part occupied by the RAF, except for one lone stag's head so far up on the wall of the Great Hall that it was obviously thought to be safe from predators. Within a few hours of the arrival of the 68 Squadron officers a cigarette had been placed in the corner of the stag's mouth. How it was done remains a mystery.

The Great Hall was ideal for social events, and dances were organised to mark special occasions, for example, the posting of 68 Squadron commanding officer, Wing Commander D. Hayley-Bell. Some lovely young nurses from the local hospital were invited to partner us, and a local jazz band provided the music. At one dance, friendly Americans stationed near the airfield provided a full dance band, which was great fun, and the rafters shook. We always took the precaution of inviting any members of the Luddington family who happened to be in the house on these occasions. This was to avoid being reported to Group for making too much noise.

By August 1944, civilian casualties caused by the flying bombs were mounting up, and further measures were needed to attempt to counter the number of bombs dropped, particularly on London. A radical change occurred when the classic method of anti-aircraft guns defending individual towns was abandoned, and these defences were distributed along the coastal strip to blaze away at anything crossing the sector. RAF balloons were similarly repositioned. Obviously this meant that both day and night fighters could not fly over the coast except through a narrow safety channel. Therefore, they had less time to get into position to intercept, which was very frustrating, and many opportunities were lost as the fighters had to break off as they approached the coast.

Unfortunately, we lost an American crew, one of three sent to fly with 68 Squadron to learn about night fighting. They followed a flying bomb into the gun strip and tragically the guns missed the bomb and shot down the Mosquito. Another of the American crews crashed at Horsted and both pilot and navigator were killed. However, I am happy to report that the third crew, John Kelly and Tom Martin, survived the war.

The British government was well aware from intelligence provided by the French Resistance movement that something like the flying bombs would be used by the Germans. As far back as 1943, 597 aircraft of RAF Bomber Command had attacked the German experimental station at Peenemünde. This attack was successful in that it delayed the introduction of the V weapons for some time. Unfortunately, forty bombers and their crews were lost on this raid. When launching sites in France were identified they were immediately attacked by the RAF, but the Germans took to using camouflaged mobile platforms, which were moved about and very difficult to find.

Fighter Command designated certain day and night fighter squadrons solely for the purpose of shooting down the flying bombs. The Tempest, then the fastest piston-engined aeroplane available to the RAF, which had only become operational two months before, was taken away from its fighter-bomber role and used to chase the bombs in daylight. Number 68, 85 and 96, all experienced night fighter squadrons, were the first to be given the task at night. The aeroplane flown by these squadrons was the Mosquito, the fastest available, equipped with radar and two cannon and four machine-guns in order to follow the flying bombs in the dark.

Apparently the total number of flying bombs destroyed by all the services remains unknown. However, records show that the RAF bag was 1,771. Intercepting the bombs by day was a comparatively simple affair once the pilot had some experience. The trick was to avoid going in too close because when the bomb blew it was liable to destroy or at least damage the attacking aeroplane.

Eventually, driven back by the Allied armies, the Luftwaffe made a last attempt to maintain the flying bomb attacks from its bases in western Germany by attaching them to their Heinkel bombers and launching them out at sea from the English coast, where they thought the Heinkels would be safe from the RAF. However, sixteen of these Heinkels were destroyed by night fighters, and claims were made for four probably destroyed and four damaged.

By Christmas 1944 the flying bomb campaign was brought to a halt, but even before this the second of the weapons of revenge, the V2 rockets, had commenced to fall on Britain. These rockets were beyond the capabilities of our defences to intercept them. All that could be done was for our bombers and fighter-bombers to attack the sites from which they were fired, but it was not until Allied armies had forced the German armies right back that the threat from these weapons was finally eliminated.

The scale of the potential German V campaign against Britain can be seen by the massive number of such armaments left unused at the end of the war. A special unit tasked to get rid of any remaining V bombs was despatched to Germany in 1945 and destroyed a total of 1,368 V1 flying bombs and 3,002 V2 rockets.

Chapter VII

'One of the freedoms of the press is that one does not have to read it.'

At the end of October 1944 No. 68 Squadron was moved again, this time back to Coltishall, but still operating against the V1 flying bombs. On the night of 19th October, Cameron and I were on patrol over the North Sea at a height of 1,000 feet, in the hope of being vectored on to a Heinkel carrier, when the port engine failed. The fire warning light came on and the propeller went into coarse pitch, but repeated efforts by Cameron to feather the propeller failed.

The propeller was turning slowly, too slowly to give power, but the drag created was causing us to lose height, which we could ill afford. Cameron put on as much power as he dared on the starboard engine, adjusted the trim and held the aeroplane straight with his right leg, which was no easy task. He was able to gain a little height, but the temperature on the good engine went up to a dangerous level, and the fire warning light came on for that engine as well.

We had already transmitted a 'May Day' call over the R/T, and we were vectored by ground control towards Woodbridge, which was an emergency airfield with three parallel flare paths, so that three aircraft could, if necessary, land simultaneously. Unfortunately, when we arrived over

Woodbridge all three flare paths were in use, and when we called the watch tower for landing instructions we were told to stand by. This was not possible as both our engines were in immediate danger of catching fire, and when we explained this to air traffic control, we were instructed to land on the 'green' flare path. Cameron did an excellent one-engine landing, just stopping short of a Lancaster bomber, which had returned from a raid badly damaged, and from which wounded airmen were being evacuated.

We were debriefed by an intelligence officer, surrounded by bomber crews telling their stories of gallant deeds over Germany. All we had to say was that we had been on patrol for one hour, with nothing to report except an engine failure! Nevertheless, we accepted our ration of rum, standard procedure at emergency airfields, and then retired to a bed for the remainder of the night. In the morning we inspected our Mosquito, and found the reduction gear casing of the port engine was so badly fractured that the whole assembly was about to fall off with the propeller. We telephoned Coltishall and Johnny Wright flew over in the station Oxford aeroplane to collect us. I have much to thank Cameron for, not least his expert handling of the aeroplane on this night, and getting us safely down at Woodbridge.

The weather over England in the autumn and early winter of 1944 was really foul; in fact in the last half of November there was no flying at all from Coltishall. In December it was also rather poor. On the 21st there was thick fog after midnight over most of the country, strangely enough except at Coltishall and other airfields in Norfolk, and thereby hangs a tale.

The Coltishall officers' mess committee had ordained that there should be a really good party at Christmas in the mess, and in view of the chronic shortage of spirits, it was deemed that the ration should be saved up, and for the two months prior to Christmas only beer should be served in the mess. However, in the early morning of 22nd December several Bomber Command Lancasters, with Canadian crews, were diverted to Coltishall because their bases in Lincolnshire were fogbound, and they remained at Coltishall for three days.

Out of the goodness of their hearts the Coltishall officers brought out their stocks of gin and Scotch for the benefit of the gallant Bomber Command crews, who drank the lot. On Christmas Eve, when the fog had lifted in Lincolnshire, they look off to fly back to their own Christmas celebrations, and the officers at Coltishall were left spiritless!

Into 1945. At Coltishall No. 68 Squadron was doing a lot of flying, with an average of 120 operational sorties every month. In January, Johnny Wright, our flight commander, was posted, and Cameron, being the senior flight lieutenant, was in command of 'B' Flight. A new edict had arrived from Group that the senior officer was always to place himself last off to fly on the night-flying list. In effect he should be present on the ground whenever possible to organise the patrols. Therefore, he and I did very little flying in January 1945, and after all that time on operations, I was not displeased.

In February the squadron was moved to RAF Wittering. Here the officers' mess seemed to be in some sort of pre-war time warp. It had an elderly civilian mess manager, dressed in black coat and striped trousers; we had the choice of boiled eggs for breakfast every day, plenty of newspapers and other nice things.

The next thing that happened was that Cameron and I were selected to attend the Night Fighter Leaders Course at RAF Ford. We spent a month on this course and we both passed. I seem to remember that Cameron received the highest rating possible. During the time we were away the squadron moved twice, back to Coltishall, and then to RAF Church Fenton, where we rejoined it.

At Coltishall, there had been more enemy activity. This and the increased number of V2 rockets coming over belied any theory that the war was about to end. On 3rd March, about seventy German bombers crossed the coast on their way to attempt to bomb the Midlands' towns. Nine Mosquitos were scrambled and there were several combats; two Junkers 188s were shot down.

These were the last combats of the war for No. 68 Squadron. Since its establishment in January 1941, the squadron had been fully operational at all times. It was credited with fifty-eight enemy aircraft, eighteen flying bombs and one E-boat confirmed destroyed, and with claims for twelve enemy aircraft and three E-boats probably destroyed, and twenty enemy aircraft damaged.

At Church Fenton on 7th April 1944, the air officer commanding arrived from Group and we were all gathered together when he made the grave mistake of announcing that the squadron would be disbanded in two weeks' time. So, of course, on the next day the first of a series of grand parties was held to celebrate this event.

It is incredible to think, when one reads in the squadron records of all the celebrations that took place, that the squadron maintained a full operational state, and even practised some formation flying for the first time! However, on 20th April No. 68 Squadron was released from readiness, and was now finished, so it was thought, as an operational unit.

The flying personnel of the squadron were dispersed to many units, some to No. 125 Squadron, some to operational training units as instructors, and the Czechs mainly to the Czech depot. It was our commanding officer, Wing Commander Bill Gill, who instructed Leslie Kohler to organise a No. 68 Squadron dinner every third week in October. This order he faithfully carried out year on year. We celebrated our 60th reunion of the No. 68 Squadron Association at the RAF Club on 23rd October 2004.

At the end of April 1945 I received my posting to the Radar Navigators School at Ouston near Newcastle together with two other chaps 'Dusty' Edwards and 'Jock' McCulloch. The chief instructor was an old friend Squadron Leader Freddie Lammer, and here, as an instructor, I expected to see out the end of my war service. No such thing! Freddie had us into his office and explained that our ideas on air interception by radar would be too old-fashioned for his school, and he would recommend that we should be posted on. This was something of a shock to me; I thought that after completing two operational tours I knew it all. Anyway, it was while we were awaiting posting that the war in Europe ended on 8th May, 1945. Dusty and Jock were seasoned drinkers and we celebrated this event in the pubs of Newcastle. Incidentally, Cameron had been posted to the Air Ministry in a desk job. Soon after this he was advised he had been 'mentioned in dispatches' in recognition of services during the war; in my opinion he deserved more than this.

By this time I had received promotion to flying officer,

which was automatic for commissioned aircrew after one year. After another year one was further promoted to flight lieutenant. It had nothing to do with particular skills, but after this the next step to squadron leader had to be worked for by promotion examinations. Anyway so it was in the peacetime RAF.

I was given a week's leave pending posting but when the notice arrived at home I was ill with diphtheria. I was stricken with a terribly sore throat and high temperature, and took to my bed. Perhaps it was really only a mild dose because the doctor did not have me isolated. The only reason I have described this illness is because of the effect it had on my next posting. I was to report to RAF Stratford (on Avon) to the Ground Control Approach Unit. Naturally I could not go, and the necessary doctor's certificate and explanation were sent there.

When it was allowed, I set off for Stratford, and was made most welcome by the station adjutant. I had missed the commencement of a course and was therefore a spare officer on the station. He, desperate to go on leave, told me I was now acting adjutant. My protests that I knew nothing about the job were ignored; he told me the Orderly Room sergeant was an excellent fellow and I could rely on him completely, and off he went. As it turned out, all I did was to attend the office during working hours, sign the things I was allowed to sign (such as leave passes), and take into the commanding officer, Wing Commander Northrop (a much decorated veteran of Bomber Command) – any problems the sergeant and I felt we could not cope with, which were very few. All the same, I was most pleased when the adjutant returned from leave, which more or less coincided with the commencement of the new course, which I joined.

Ground Control Approach (GCA), in simple terms, was a system of instructing pilots how to land in bad weather, particularly fog. Indeed, I believe it is still in use in a modern form on military and civilian aerodromes. This was, of course, radar again, and June 1945 marked my fourth year involved in this wonderful invention. The first shock was when we were voice tested, and on the playback I found my voice high pitched and squeaky. With others on the course, I had to take lessons in voice production at the proper level. As you can imagine, it is essential to have the right sort of voice over the R/T to give a pilot confidence when you are talking him down in bad weather. Another thing of the greatest significance was to keep the voice steady whatever the pilot was doing in response to your instructions. Student pilots, we were told, were particularly sensitive to the voice of the operator talking them down. Any sudden change in inflection in the GCA operator's voice could cause unnecessary panic, which was, of course, best avoided.

I cannot recall the details of the Ground Control Approach course (and wouldn't bore you with them if I could), but it must have been pretty tough because of the responsibility of giving the right instructions to incoming pilots in inclement weather, which could mean life or death for them. However, watching a radar screen had been my job for the last three and a half years, and I had gained a certain expertise in giving instructions to pilots during that time. I found the instructors were first class at Stratford, and once I had got the hang of things, I felt I could deal with talking pilots down to the runway with confidence.

However, I was still very surprised that when the course was finished I was made an instructor at Stratford. All the other pupils on my course were posted away to work as operators at various RAF units, many to Germany, where the RAF was taking over the airfields previously used by the Luftwaffe.

Then followed fourteen months of most enjoyable service at Stratford. I had some great companions there. All the other instructors were ex-aircrew who, like me, had completed a tour or two on operations, and were now tour expired. We spent much of out time as instructors in the trucks at the end of the runway. In each truck were two screens enabling two pupils at a time to be taught. There was, however, a time limit of three hours at this task by RAF law (for both pupils and instructors). Therefore, we spent the rest of the time in the classrooms, giving lectures on various aspects of the course. It must have been a dead bore for the tutor pilots, constantly landing and taking off – talk about circuits and bumps!

In spite of the amount of work detailed above there was plenty of time off, and if I was in the truck instructing in the morning it usually meant I had the afternoon off. Stratford-upon-Avon golf club started up again post-war, actually in the spring of 1945, and I was persuaded to go round with my room-mate Bill Pettifer. It was very good of the club to make all the officers at Stratford airfield non-fee-paying members, and after a few lessons from Bill I was beginning to enjoy the game. Then Bill was posted away, and when he left he offered me his bag of seven clubs for £7. I took up his offer and then continued to play almost every day of good weather.

I remember that one of the clubs was an old-fashioned driver with a metal head, and one day in the mess the station commanding officer looked in and asked whether anybody could lend him a bag of clubs for the day. Like an idiot I volunteered. When they were returned, the driver was broken, snapped off at the head. I have to report that the CO did not offer to pay for the broken club!

We really had a very splendid time in Stratford-upon-Avon. The theatre reopened for the winter season in 1945, not at first for Shakespeare, but for other plays, and the town again became a mecca for the tourist. The number of young women about was a constant source of companionship for us chaps. I remember the pubs of Stratford were particularly good. The Swan's Nest, going into the town from the airfield, just before the bridge over the Avon, was our favourite. If one did not particularly wish to go out, there was usually a poker school going in the mess. I remember I was £80 up at one time, but I expect I lost it all later.

Now, I must explain a set of circumstances that illustrates the darker side of my nature, or, to put it another way, I am not as nice as you might think. Early in 1945, I decided that during a leave I would visit my company, the South African Wine Farmers Association (London) Ltd, situated at Nine Elms, in south-west London, where we had our own bonded warehouse and bottling plant, as well as the offices. I thought it best to keep in touch because demobilisation would occur for me within the next year. Also, I now had two bands on my sleeve, as a flight lieutenant, with which to impress them.

I had a long talk with Mr Craig, the general manager, and, as expected, the main subject was when would I be coming back to rejoin the company. I explained that as an unmarried fairly young man who had not joined the services until December 1940, and as the war was not long over in the Far East, I was not likely to be released for some time. He then asked me whether he could do anything to hasten my demobilisation because shipments of wine were already on their way from the Cape after a five-year break, and he needed me to help cope with the expected rush of orders when the UK wine merchants replenished their stocks. I must say I was quite flattered to hear this, and, of course, relieved that I was going to take up civilian life again at the company in a good position. So what could I do but suggest he wrote to the commanding officer at Stratford.

I had hardly returned from leave when I was called into the office of the chief instructor, Squadron Leader Rex Brailsford, who showed me Mr Craig's letter. He knew, of course, that we were on to a good thing at Stratford, and were enjoying ourselves, because he just said, 'Do you really want to go back to your company now, or would you rather stay here until demobilisation?' SAWFA had been paying me throughout the long years of the war, they were a very good company, my future depended on them, and yet given the chance to help them I replied, 'I would rather stay here for as long as possible.' On the recommendation of the chief instructor, the commanding officer wrote back to Mr Craig saying I could not be spared, and I stayed on at Stratford until I was demobilised in August 1946.

That decision has always been on my conscience, and I really do not deserve to have had employment in such a good company as the South African Wine Farmers Association.

I have mentioned previously that the company was, very kindly, paying us fifty per cent of our salaries on call-up throughout our service in the forces, half each week and the balance when we returned. So when I got back I received a cheque from SAWFA for £275, which was much larger than my RAF demobilisation gratuity as a flight lieutenant of £109 19s 6d!